Martin is a writer, surfer and beach lover. He founded the Beach Clean Network with Tab Parry in 2009 and started the #2minutebeachclean hashtag in 2013 after North Atlantic storms left UK beaches littered with plastic rubbish. It's a simple, effective idea – pick up beach litter for 2 minutes, bag it, tag it, bin it – and the hashtag has been used many thousands of time around the world. *No.More.Plastic.* continues the clean-up beyond our beaches with a #2minutesolution for everyone.

A percentage of profits from the sale of this book go to The Beach Clean Network. Thanks for your help.

NO. MORE. PLASTIC.

What you can do to make a difference

MARTIN DOREY
Founder of the #2minutesolution

EBURY PRESS

This is for the barefoot family of #2minutebeachclean warriors.

And anyone, anywhere, who ever went out of their way to pick up litter.

Heroes, one and all.

16

Published in 2018 by Ebury Press an imprint of Ebury Publishing,

20 Vauxhall Bridge Road,
London SW1V 2SA

Ebury Press is part of the Penguin Random House group of companies
whose addresses can be found at global.penguinrandomhouse.com

First published by Ebury Press in 2018

www.penguin.co.uk

A CIP catalogue record for this book is available from the British Library

ISBN 978 1 785 03987 4

Typeset by seagulls.net
Printed and bound in Great Britain by Clays Ltd, Elcograf S.p.A

Penguin Random House is committed to a sustainable future
for our business, our readers and our planet. This book is made
from Forest Stewardship Council® certified paper.

CONTENTS

HAVE YOU
GOT 2
MINUTES?

OF COURSE
YOU HAVE.

FOREWORD

Rock, scissors, paper. Rock, scissors, paper ... plastic.

When we were kids we'd sometimes add dynamite, the killer trump, but plastic, that was the material, the tool, the toxin to beat them all. And now, at last, we all know that's true, so there is no escaping the job of reducing, re-using and recycling and we must get on with it. And, if you need a push, read on.

I like Martin Dorey. He's a nice bloke but what makes him special is that he's also a real doer. I also like simple ideas that work like his #2minutebeachclean and, thankfully, so do thousands of other people because together they have harvested tonnes of plastic waste from our beaches. It's an ambitious initiative that's all about connecting with people, being reasonable about their capabilities and making them a part of something positive. Beach cleaning is a cure, it helps sort out the mess we've already made. But wouldn't it be much easier if we prevented making the mess in the first place?

What we need is the simplicity of the #2minutebeachclean, something that we can all do, without too much pain or cost in our everyday lives, which will really make a difference ...

It's here, in these pages: the #2minutesolution. I read it yesterday and I've done three things today and that's testament to Martin's brilliant vision and ideas. Now it's your turn!

Chris Packham

(A man with a new water purifier, packaging-free onions and degradable dog-poo bags)

THIS
BOOK
IS
ABOUT
YOU

This book is also about plastic and the massive problems we are facing with plastic pollution.

The reason it's about you is because you are the one person in the world that you inhabit who can make a difference. You really are.

I guess you already know we face a big problem from plastic pollution. We are drowning in the stuff. Our oceans are choking. Birds, fish, cetaceans and marine mammals are dying in their hundreds of thousands each year because of plastic. It strangles and entangles them or they mistake toxic broken fragments of it for food and die: miserably, starving to death with full stomachs. The legacy of almost 100 years of out-of-control plastic production is coming back to bite us. And now, suddenly, we are waking up to it.

Don't worry – if you think I am going to tell you off for using plastic in your everyday life, I'm not. Quite the opposite, in fact, because I am going to thank you for every little bit you do.

Plastic is a wonder material with all kinds of brilliant uses. But we need to use less of it, use it more wisely and stop letting it pollute our oceans.

We might look to inspiring individuals or memes on Facebook to solve the problem for us, but they can only do so much. Sharing their stories might spread the word and make you feel like you did your bit, but it is not enough to make real change happen.

It's the same with governments or corporations. They are concerned with business and the business of being popular, with little appetite to act decisively, especially if it means passing unpopular laws or damaging the bottom line.

The solution, therefore, is down to you and me.

So this book is about what you can do to make a difference, why it matters, and how it won't take you a lifetime to do it. In its pages you'll find ideas for reducing your plastic consumption. Each idea takes the form of a #2minutesolution – a simple, easy way to give up plastic. While each #2minutesolution might be nothing in itself, they all add up. A #2minutesolution will take you no time at all – 2 minutes! – and yet will be a step towards making a meaningful difference to the world.

I'll also tell you about a few inspiring ideas that have spread to such an extent that they are already making a huge difference. They all started with just a spark.

That's all it takes to light the fire.

What you can do now

Every chapter in this book has a section like this. These are things I'd like you to do or think about. It may be as simple as reassessing the plastic in your life, taking a look at the plastic in your local supermarket or challenging you to pick up plastic for a few minutes, just to see how much you can find. The idea is that each of these will add up. When you get to the end of the book (it won't take you long) you'll see how much you can do – without doing much at all!

WHO
AM I
TO TELL
YOU
THIS?

In 2009 I moved to a small hamlet near a quiet beach on the north Devon coast. One day I came across an area of the foreshore that was knee deep in plastic bottles. I vowed to do something, anything, to clear it up.

I decided to act and, with my friend and colleague Tab Parry, formed the Beach Clean Network, an organisation that was devoted to putting beach clean organisers in touch with volunteers.

Later, in 2013, after a series of fierce Atlantic storms, I was completely overwhelmed by the plastic the ocean spewed up on the beach. I knew I could never pick it all up alone, so I turned to social media for help, creating the first hashtagged #2minutebeachclean post. The idea was to try to inspire people to do the same – pick up litter for 2 minutes –so that it became a part of life. In the beginning I figured that if I inspired just one other person to do the same, then it would double my efforts – and that would be a win!

Since then the idea has blossomed and people have responded in their thousands. Each week around 145,000 people look at the @2minutebeachclean Instagram account, while stats tell us we often get over 150,000 views of our tweets each day. In the week John John Florence (WSL world champion surfer) posted about us, over 350,000 people looked at our Instagram account.

At the time of going to print there were 64,706 posts to Instagram using the hashtag, from every continent, including Antarctica, and it's rising by around 100 every day.

Add to that the countless other posts to Twitter and Facebook, those that have been spelled wrong (there are a lot!) and all the people who do their #2minutebeachclean and don't share it on social media. We estimate

INSTAGRAM TIME:
If we add up the time spent picking up litter based on the number of Instagram posts at the time of going to print, it amounts to over 12 weeks of clean-up time.

that the rubbish collected during each #2minutebeachclean weighs about 2 kilos, which means that the weight of litter removed is at least 130 tonnes, probably a lot more.

After the BBC's flagship nature documentary *Blue Planet II* aired in 2017, showing the horrific scale of the plastic pollution problem in our oceans, the programme's website suggested ways people can get involved in ocean conservation. The #2minutebeachclean was top of the list.

In 2015, with the help of surf retail giant Surfdome and Keep Britain Tidy's BeachCare scheme, we launched 8 trial 'beach clean stations' in Cornwall. They are specially designed 'A' boards that hold litter pickers and bags to make it easy for beach-goers to do a #2minutebeachclean. After a year, we were able to look at litter data from Crooklets Beach in Bude, and discovered that there had been a 61 per cent drop in litter during the trial period.

We now have over 350 stations in the UK and Ireland. They are proving really popular and are good for tourism – as they help beaches qualify for Blue Flag Awards.

And that brings us to now.

It all started with 2 minutes.

What you can do now

We are so used to accepting plastic in our lives without thinking. It's become normal for everything to be packaged in plastic. And yet each piece of plastic we throw away will not go away. So it's time to think about the consequences of using – and throwing away – so much plastic. If you do one thing today, think about the plastic in your life. Look around you. How much do you see?

THE #2MINUTE-SOLUTION

Ocean plastic is nothing more than a symptom of our overconsumption of plastic, which means the #2minutebeachclean is a sticking plaster. While any effort to clean up the ocean is vital, we have to stop the tide of plastic at source.

The #2minutesolution is our way of doing it.

Each #2minutesolution is a simple way of quitting the plastic habit. It's when you add them up that you start to make a difference.

What you can do now

Our love of convenience has been leading us a merry dance. It's been too easy to have it all and have it all now. Sadly, the consequences are coming back to bite us – all that packaging! *Grrrr!*

Think of one item you buy in single-serve or handy packs – like tissues, yoghurts or individually wrapped crackers and dips – that's about your convenience more than the planet. Cut it out. Add up the amount of packaging you'll save from going into landfill (or the ocean) after a year.

WHY 2 MINUTES IS THE KEY

Two minutes is shorthand for 'no time at all'; a perfect chunk of time that's no hassle. It's nothing to take 2 minutes out of your day to do something positive. Making the decision to do something is made easy because it's no bother. It's time you can spare.

It's only once you add up all the 2 minutes together that you get to an hour, a day, a week. Before long, you've achieved something without making much effort at all. Two minutes isn't too much for your lazy days and won't get in the way of your busy life. It lets you know when it's OK to stop and when it's OK to stop feeling guilty about not doing more.

That's the key.

All you have to do is take the first step.

What you can do now

Go outside, into your street, a local park or open space. Set the timer on your phone to 2 minutes and pick up litter until the buzzer goes off. How much did you collect? Surprised?

WHY NO MORE PLASTIC?

Our world – and every living creature in it – is under threat from plastic. It's in the ocean as well as along our verges, on our streets, in our parks, on the beaches, up the mountains, in lakes, rivers and streams. You can't go anywhere without seeing plastic.

If we don't stop the flow, the ocean is where it ends up. As a result, every tide brings plastic to our shores, while the 'gyres', huge areas where microplastics – tiny fragments of plastic – are suspended in the water column, are now present in every one of the world's oceans.

The Great Pacific Garbage Patch is estimated to cover an area extending to as much as 1,500,000 square kilometres. That's more than 60 times the size of the UK's land mass.

If we don't change, the plastic problem is only going to get worse.

In 2010, *National Geographic* reported that 8 million tonnes of trash enters the world's oceans each year.

In 2017, the Ellen MacArthur Foundation estimated that by 2050 there would be more plastic in the oceans than fish.

What you can do now

Do you hate blister packs? Those clear plastic packs that everyday items like scissors, toothbrushes, pens and batteries come in? Me too. They are so unnecessary and are often more about merchandising – the way shops display products – than keeping the product pristine.

Think of the last thing you bought that came in a blister pack. Can you buy it without?

Next time you need it, vote for the product with the least amount of packaging.

WHAT'S THE PROBLEM WITH PLASTIC?

Plastic, in short, is poison.

Hailed as a wonder material when the first synthetic polymer, Bakelite, was developed using fossil fuels in 1907, plastic has made so much possible, from the way we manufacture to making our lives faster and more convenient.

However, its usage as a single-use product – and our reliance on it, as well as our failure to recycle or reuse it properly – has made it a nightmare material. This is because plastic is persistent. It does not biodegrade and disappear. It becomes brittle over time and degrades into smaller and smaller pieces, known as microplastic. Every piece of plastic ever made is still around, and may well be for the next few thousand years, either in the ocean or as a toxic time bomb in landfill.

Plastic leaches chemicals, such as BPA, which mimics oestrogen, and has been linked to low sperm counts and infertility in men, as well as breast and prostate cancer. In the ocean, plastic also attracts persistent organic pollutants, which are naturally occurring toxins. These accumulate over time, meaning that any ingested pieces pass the toxins on to the creature that has eaten them. This can travel up the food chain to us.

That's bad news.

Oh, and there's another thing: plastic is made from fossil fuels.

What you can do now

Go to the veg aisle at your local supermarket. Count up the number of items that could easily be sold without plastic. Do you see what I see?

It's only once you start to look that you can really start to see.

WHERE DO MARINE PLASTICS COME FROM?

Plastic gets into the ocean in many ways. It washes down rivers from cities, blows into the sea from inland, is lost or dropped overboard from ships. A plastic bottle, thrown from a car or left on the street, can find its way to the sea down a storm drain. Some plastic finds its way down the sewage system – plastic tampon applicators and cotton-bud sticks, for example – when it has been flushed down toilets miles from the sea. Some plastic gets washed overboard ships in shipping containers that then burst open and spill their contents. Some plastics are lost from fishing boats or washed out of rubbish dumps. Other plastics – raw-material plastics in pellet form – get released accidentally from plastics factories.

It's easy to blame someone else for plastic in the ocean. Some argue that it comes from Asia, or shipping, or the USA (a lot of litter found on beaches in the south west of the UK comes across the Atlantic on the Gulf Stream), or that careless louts leave it on beaches and, therefore, it's got nothing to do with them.

But the basic fact is that 100 per cent of marine litter comes from us. We are human and humans make plastic.

It's ours. And we have to clear it up.

What you can do now

Spend a couple of minutes going through your supermarket shop. Look at the packaging. If it says 'cannot currently be recycled', don't buy it again.

WHY YOUR 2 MINUTES MATTER

The problem with ocean plastics is daunting, isn't it? It's easy to feel helpless in the face of so much plastic rubbish – around 8 million tonnes – that finds its way into the oceans each year.

What could you possibly do that will make a difference?

The thing to remember is that everything you do affects the world in some way.

We always tell anyone who ever does their #2minutebeachclean that it matters because every piece of plastic that gets removed from the ocean environment, or any environment, is a piece that won't go on to kill. It won't end up in a whale's stomach or strangling a seabird. And it won't end up becoming thousands of toxic pieces of microplastic.

Everything you do has an effect. Every piece of single-use plastic you refuse is a piece that won't end up in the ocean. Every letter you write gets noticed. Every protest you make gets heard somewhere.

All those actions add up. That's why I know we can change the world, 2 minutes at a time.

What you can do now

Next time you go out, pick up the first plastic bottle you see and take it home to recycle. How long did it take? And yet that's 1,000 pieces of microplastic not floating around in the ocean in years to come.

PLASTIC ALTER-NATIVES

We are developing alternatives to plastic all the time. But it's not that simple. Just because something says it's compostable or biodegradable, doesn't mean you can use it without thinking about where it goes after you've finished with it. And if a bag is described as 'degradable', like dog-poo bags, it doesn't mean it's OK to leave it on your dog walk.

Degradable, oxo-degradable and photo-degradable plastics are still plastic. They just break down into smaller and smaller pieces.

Biodegradable means something can be broken down into organic materials without causing harm or leaving toxins behind, in a reasonable timescale. Some biodegradable products can only be degraded under controlled conditions, in a commercial composter.

Compostable products break down into nutrients that enhance the soil, although they may have to be disposed of in a commercial composter, rather than at home.

Home compostable is exactly that. You can put it in your home compost. Check how long it will take to break down before you use it! It may take months.

Bioplastic, made from naturally occurring substances, could be the future. If edible or beneficial to the environment then they could be the breakthrough we need – as long as they don't cause problems if they degrade in the environment. If they degrade at all.

Recently a plastic toy manufacturer – that produces around 20 billion pieces of plastic a year – announced it would be making a 'sustainable' range out of bioplastic made from sugar cane, a sustainable resource. They do not say how long it will take to degrade at the end of its useful life.

What you can do now

Look out for misleading claims on packaging. Plastic is still plastic, even if it is 'degradable'. Some 'eco-friendly' packaging can still only be composted in industrial composters and will spoil recycling streams if it gets into them. While sustainable plastics are better than oil-based, it's still important to think about what will happen to it after you've used it.

IS RECYCLING THE ANSWER?

Recycling is essential, of course, because we have finite resources and we have to use them wisely. No one is going to live on another planet anytime soon.

Most types of plastic can be recycled, but the world of recycling can be confusing because one local authority may not be able to accept the same plastics as another. So it's different all over the country. This is to do with local contracts, market value and how easy or financially viable it is to recover material. Some items are made of composite materials, such as ball-point pens or Tetra Paks, which make it difficult to recover the raw materials. Other items, like bottles with plastic wraps, cannot be identified by recycling sorting machines and get rejected, so going straight to landfill. Also, some eco-materials cannot be recycled within the normal recycling stream and can spoil whole batches of material, so condemning the whole lot to landfill.

It's a minefield.

However, I still believe we have to recycle as much as we can. Aside from the confusion it brings, the problem I have with recycling is that you and I pay for it.

Recycling is a business and recyclers get paid to take away our litter so they can sell it on. But it's the local authorities who pay the recyclers to provide that service.

The people who make the plastic don't pay. Does that seem right to you?

Sadly, putting plastic bottles in your recycling bin does not always guarantee that it'll get recycled either. It may get used for energy recovery (burnt) or it may end up in a spoiled batch or being sent to another country to be recycled, where it may get rejected and sent back or put into landfill.

The only way to guarantee its future is to do without it.

Identifying plastic types and their usage

Manufacturers use a rating system to help make identification easier for recycling. Plastics are classified into 7 groups, each of which has different characteristics and usage.

CODE AND SYMBOL	PLASTIC TYPE	TYPICALLY USED FOR	PROPERTIES
01 PET	Polyethylene terephthalate	Soft drinks bottles, food trays	RECYCLABLE Clear, tough, sinks
02 HDPE	High-density polyethylene	Yoghurt containers, shopping bags, milk bottles, shampoo, detergent and chemical bottles	RECYCLABLE Floats in water
03 PVC	Polyvinyl chloride	Blister packs, pipes and hose, clear food packaging	RECYCLABLE (sometimes) Considered to be the most toxic of all plastics. Not recommended for food use.

CODE AND SYMBOL	PLASTIC TYPE	TYPICALLY USED FOR	PROPERTIES
04 LDPE	Low-density polyethylene	Rubbish bags, squeezable bottles, clingfilm	RECYCLABLE Floats in water
05 PP	Polypropylene	Bottle caps, straws, food tubs	RECYCLABLE Floats in water
06 PS 06 PS-E	Polystyrene, expanded polystyrene	Plastic cutlery, video cases, CD cases, cups, plates	NOT EASY TO RECYCLE Leaches styrene, a suspected carcinogen. Not recommended for food use.
07 PC, OTHER	Polycarbonate resins and composite materials	Components, computers, electronics	NOT EASY TO RECYCLE The 'catch-all' for any plastics that can't be categorised by the other 6. Not recommended for food use. Risk of containing BPA.

What you can do now

Find out what your local authority can take for recycling and where it goes. Make a pledge to live within what they can accept. If they don't accept some materials, ask them why.

THIS BOOK IS NOT ABOUT ...

POLITICS, ECONOMICS OR ADVERTISING

However, there are connections between our plastic crisis and politics, economics and advertising.

Firstly, advertising has played a big part in defining our throwaway culture and the way we consume. Advertisers make products look irresistible, often generating demand by creating anxieties where none exist. Glossy ads imply you can be sexy, thin, funny, rich or 'living the dream' if you buy their products. Until recently, very little thought was given to what happens to the packaging or product at the end of its life with you. Can it be recycled? Is it made from sustainable materials? Does it go to landfill?

Secondly, our society, since the mid-fifties, has been about convenience. Our politics has given us 'market forces' and freedom for corporations to sell their products with impunity. Our focus has been on growth, business, trade and progress. And in that climate, Plastic PLC has thrived unchecked.

Now Earth needs another way and it's up to us to make it happen. We can't wait around for politics because politicians follow what is popular.

We need to turn to the simplest of economics.

Market forces tell us that if we stop buying plastic, the producer will have to change or go out of business. So if a product doesn't fit your ethos, don't spend your money on it.

Vote with your wallet.

What you can do now

Email your MP. Tell them you want anti-plastic legislation NOW, that you want polluters to pay NOW, that you want your children to grow up in a plastic-free world NOW. It is their job to listen (especially if they want your vote).

In the UK, find your MP at:

www.theyworkforyou.com

A template email to send to your MP:

Dear [INSERT YOUR MP'S NAME HERE]

I am writing to you to express my concern at the scale of plastic pollution in our oceans. While I am doing what I can on a personal level to avoid single-use plastic, I can only do so much. So I would like to call on you and Parliament to introduce tough legislation to tackle the plastic crisis now and not in 25 years' time.

I would like you to force polluters to pay for the plastic clean-up, to impose high levies on corporations that package their products in unnecessary plastics and to ban non-recyclable plastics from being used in everyday packaging.

I want my children – and every child – to grow up in a world where plastic isn't contaminating our oceans and waterways, killing animals and causing a risk to our health.

Please help make this happen.

Yours,

[YOUR NAME HERE]

THE 7 WORST OFFENDERS

We released the #2minutebeachclean app in January 2018, to help beach cleaners in the UK and Europe count what they find. They can use it to take a picture of their #2minutebeachclean, then log the amount of each type of litter they have picked up. They can also attribute their #2minutebeachclean so, in time, we'll be able to see patterns of what type of litter is washing up where.

Already, after more than 35,000 pieces of litter have been logged at more than 800 locations all over the world, it is helping us to see what's washing up.

Opposite is a list of the top 15 found items. From this list, we can see that 7 of them are items that we all use every day. These 7 items (marked in bold) make up almost a third of all the items that have been logged, so cutting them out of your life altogether – or finding non-plastic alternatives – will make a massive difference.

1. 9.97% Microplastics
2. 9.03% Bottle tops/lids
3. 8.58% Net pieces
4. 7.02% Polystyrene
5. 6.95% Hard plastic pieces
6. 6.01% Drinks bottles
7. 5.78% Rope
8. 4.54% Nurdle
9. 4.25% Drinking straws/plastic cutlery
10. 3.89% Crisp packets/sweet wrappers
11. 3.7% Q-tips/cotton-bud sticks
12. 3.67% Fishing line
13. 3.25% Plastic bags
14. 1.81% Wet wipes/pads
15. 1.65% Foam

What you can do now

Take a look at the list on the previous page. Which 3 out of the 7 worst offenders could you give up tomorrow? Do it.

HOW YOU CAN CUT THEM OUT

When you look at the figures for what's washing up on the beaches it's easy to see that there are things you can do today to reduce your plastic impact that will have an effect on what's in the oceans. Here's the background on the Top 7:

❶ AND ❷ BOTTLES AND BOTTLE TOPS

This is your first #2minutesolution! And it's possibly the easiest to do.

According to Recycle Now, every UK household uses around 480 plastic bottles each year, but only recycles around 270.

Do the maths and you'll soon get the picture: over 35 million plastic bottles are used every day in the UK and around 23 million of them won't get recycled. Where do they go? Either to landfill or into the environment. They don't 'go away', because, to coin a phrase that's used a lot these days, there is no 'away'.

Why do we find more bottle tops than bottles? Plastics can be identified by their ability to float or sink in water. Plastic water bottles, made from PET (type 1), will not float unless they have a lid on and air trapped inside. The type of plastic used for lids, HDPE or LDPE (typically type 2 or type 4 plastics) will float, which is one of the reasons why people using our app find more bottle tops than bottles. PET bottles without lids go straight to the ocean floor, where they risk being broken up into thousands of pieces of microplastic.

Why buy bottled water?

3,200,000,000 litres of bottled water (according to the Bottled Water Association) are sold in the UK each year. In the supermarket, water costs anywhere between 30p and 60p per litre, which means that if you buy a litre of bottled water a day in the UK you could be spending up to £220 a year on something that's available from a tap. And you'll pay around 500 times more for it in bottled form. Big scam? Probably.

Bottled water comes in plastic, usually made from virgin plastic (non-recycled), which is made from oil and has to be transported (with a high carbon footprint). It sits on a shelf until you buy it, where it may leach chemicals – such as BPA and dioxins as well as microplastics – into the water. And while it has to pass safety standards, it is only tested when it is bottled. Recent studies also showed that 93% of bottled water showed signs of microplastic contamination.

The #2minutesolution

So here it is: use the tap. In the UK, tap water is some of the best on the planet. When tested and measured for compliance to standards set by the EU, British tap water was on average 99.9 per cent compliant. It also costs about 1p per litre. So why waste your money on bottled water?

If you worry about impurities, chemicals and minerals in your tap water, you can buy a water filter (around £20) that will 'clean up' your tap water for a fraction of the cost of bottled water.

Getting a free refill

The Refill scheme, which runs in lots of British cities, started in my home town of Bude as a way of raising money for the local sea pool. Local cafés offer tap water free to anyone with a refillable bottle from the sea-pool shop. It worked so well that it was taken to Bristol, where Natalie Fee, an anti-plastic activist, brought it to life. Refill now exists all over the UK, with an app that tells you where you can refill for free.

As an example of how a simple thought can change the world, this is the finest. It has grown into a campaign with clout. It only takes seconds to refill a bottle, saves you money and prevents single-use plastic water bottles from going to landfill or the environment.

No excuses, right?

What you can do now

Download the Refill app now, find your nearest free refill point and use it.

It is estimated that Refill has prevented over 2.1 million single-use plastic bottles from being used in the UK in just 3 years.

❸ DRINKING STRAWS AND PLASTIC CUTLERY

Drinking straws

From the stats collected from the #2minutebeachclean app, around 4.25 per cent of the litter comprised of plastic straws and cutlery. While 4 per cent might not sound like much, it adds up to a lot on a global scale. It is estimated that around 500 million plastic straws are used and thrown away each day in the US.

Have you seen the video of the turtle having a straw removed from its nose? Do you need more proof that straws suck?

In 2017, the UK pub chain, Wetherspoons, vowed to cut out plastic straws. This will save 70 million a year from either going to landfill, or making their way into rivers streams and, ultimately, the ocean.

The #2minutesolution

Refuse straws whenever you get offered them. You might have to be vigilant because bartenders and waiters are pretty nifty at putting them into your drink without you asking. There's no point in getting annoyed, but if you get a chance to chat, tell them why.

Better still, frequent places that don't give them out.

What you can do now

Be brave. Next time you are out, refuse the straw, tell them why and be proud.

Plastic cutlery

Along with straws, plastic cutlery is one of the top offenders. And yet it's so easy to avoid. Thankfully, we are waking up to it, asking food outlets, governments and big companies to cut it out altogether. We are making changes. But there is still more to do. We need to challenge ourselves and others to stop using single-use plastics like cutlery and find effective alternatives.

We need to free ourselves from the tyranny of convenience.

There are about 7.5 million children at school in England. Just imagine if each one of them uses 1 plastic fork each day for the whole of the school year. That's 1,425,000,000 plastic forks that could be avoided by using metal cutlery.

I'm not saying that all schools use single-use plastic cutlery, but I know that some do.

Think about it. Now think about all the sandwich shops, takeaways, supermarkets and shops that are giving you plastic cutlery in a plastic bag each time you have a takeaway lunch or dinner.

The #2minutesolution

No meeting, phone call, email or chat with the boss is important enough that you can't find 5 minutes to use a proper fork! So try to do without that plastic one – and don't be afraid to say so next time you get your lunch.

What you can do now

Go to your kitchen drawer. Take out a fork. Put it in your work bag. Use it. Think how much of a difference it would make if we all did it!

❹ CRISPS AND SWEETS

'Not currently recycled' is the message I keep seeing on sweet, crisp, biscuit and nut packets. As far as I am concerned this should be illegal. How can anyone justify making packaging out of plastic that will only end up in landfill, the ocean or in an incinerator? It is not unusual to spot sweet wrappers from twenty years ago washed up on beaches. One recent example was a Marathon wrapper, which looked in great condition. Marathon became Snickers on 19 July 1990, which means that wrapper was floating about for at least 28 years. This goes to show how persistent plastic is. And plastic from sweets and crisp wrappers is often made from composites of plastics, making them really expensive and difficult to recycle.

The #2minutesolution

Buy big bags and deal them out if you or your family can't live without crisps – it uses much less non-recyclable plastic than six small bags inside one large one.

Buy chocolate and sweets that are wrapped in foil and paper or seek out a traditional sweet shop or pick 'n' mix that sells sweets in bags.

The UK's biggest producer of potato crisps, churns out 10 million packets of crisps a day. Sadly the packets are 'not currently recyclable'.

What you can do now

Go and take a look in the cupboard. Pull out the items that have 'not currently recycled' on the packaging. Think about how you could live without them. Seek out alternatives.

5 Q-TIPS AND COTTON-BUD STICKS

Cotton buds are used by millions around the world. While we know that they should never be flushed, it is inevitable that some people will lose a few down the toilet. Oops.

Once in the sewage system, they can't always be stopped by filters. This means they end up in waterways and, ultimately, in the ocean. If you attend any beach clean you will find cotton-bud sticks, or pieces of cotton-bud sticks.

The #2minutesolution

In 2016, a UK campaign called 'Switch the Stick' collected over 150,000 signatures for a petition to major retailers to change from plastic to paper. At the end of the campaign, UK retail chains Tesco, Sainsbury's, Asda, Morrisons, Aldi, Lidl, Superdrug, Boots UK and Wilko pledged to switch to cardboard sticks.

You can buy cotton buds now knowing they won't pollute the oceans. Just check before you buy that they are made of 100 per cent cotton and paper.

What you can do now

Undertake a 2-minute review of your toiletries. Start with cotton buds and move on from there. If there is plastic in them, think about how you can make changes to cut it out.

Cotton-bud sticks make up 3.7 per cent of the total number of items logged on beach cleans using the #2minutebeachclean app on beaches across the world

6 PLASTIC BAGS

The 5p Bag Tax on plastic shopping bags had a huge effect in the UK. Since the tax was introduced, the number of bags given out by supermarkets has decreased by more than 80 per cent.

Plastic bags are deadly to wildlife, particularly sea mammals and turtles. Turtles mistake them for their favourite food – jellyfish – and eat them. Since they cannot be digested, the bags sit in the animal's stomach, making them unable to feed. Eventually, they die.

The #2minutesolution

Cut out plastic bags for ever. You can buy cotton bags for life from a huge variety of places now, many supporting good causes, including at www.beachclean.net.

You can get free bags, called Morsbags, from local bagging groups, which are made out of offcuts of material and then given away. They come in all shapes and sizes and every one is different. You can find your local Morsbag group at www.morsbags.com. They are all over the world. Join them!

What you can do now

Make a Morsbag yourself. Download the pattern, buy your labels and start making. It won't take long to make one. Then you'll be off and running ...

Over 260,000 Morsbags have been made since the idea was set up in 2007 by Claire and Jo Morsman.

7 WET WIPES AND PADS

Wet wipes are awful.

Sorry.

Their construction – often they are made from plastic fibres – means that they don't break down like loo paper, so while they are excellent for wiping and washing in extreme situations, they are a victim of their own success. Water companies hate wet wipes. They clog up sewer pipes, getting entangled in fatbergs (giant lumps of cold, coagulated fat that gather in sewer pipes) and in the worst cases causing flooding. In the ocean, they take ages to break down into microfibres, but not before they risk being eaten by sea creatures.

While manufacturers put 'don't flush' icons on plenty of wipes, we obviously can't be trusted with them. So it's time to give them up.

It's the same for a lot of incontinence and menstrual products. They often contain synthetic fibres, which means they don't break down in the ocean or landfill. So they end up on the beaches. Yuck!

The Marine Conservation Society has seen a 400 per cent increase in wet wipes found along the UK coastline over the past decade.

The #2minutesolution

Wet wipes are a fine example of personal convenience over good sense, especially given what we now know about the oceans. It's easy to buy a reusable flannel or use paper tissues. If you are taking off makeup, try using cotton-wool pads or bamboo wipes, both of which can be composted.

If you are changing nappies and can't do without them, don't panic. Use damp paper towels, cotton wool, bamboo wipes or cloths that can be washed. If you MUST keep using them please, please, please remember to NOT TO FLUSH THEM.

What you can do now

I dare you to Google 'wet wipes sewage' and look at the images. Enough said?

#2MINUTE-SOLUTION FOR YOUR HOME

There are lots of ways in which you can reduce your plastic consumption at home. So many products we buy come in plastic or are made from plastic. Cutting out any or all of them will have a twofold effect: firstly, you'll be using less, which means less going to landfill or recycling; secondly, you'll be sending a message to manufacturers that you don't want products made like that any more. It might not seem like it matters, but it does, because the more of us who take a stand, the more we get done.

MEN'S SHAVING PRODUCTS

While disposable razors (with toothbrushes, combs and pens) make up less than a half of 1 per cent of items picked up and logged using the #2minutebeachclean app, they play a very relevant part in this story. They are a prime example of the way that advertising affects society and makes us believe we aren't good enough unless we keep up with 'progress'.

Shaving is something that many of us do, so it's ripe for exploitation. In order to keep selling more, manufacturers have reinvented the razor many times over. And as a man – or woman – you are made to feel that only 3 blades will do. Oh, no, hang on, only five will do. And so it goes on.

It is estimated (figures from statistics portal, Statista) that the global male-grooming market will be worth US$29.1 billion by 2024. That's a lot of money on products that are, by their nature, throwaway.

Even non-disposable razors with replaceable catridges are thrown away every few weeks, inevitably ending up in the ocean or in landfill.

The #2minutesolution

The easiest way to reduce your shaving waste is to use a safety razor. You'll save money too.

I bought a safety razor, which uses real razor blades, for £4.99 and a pack of blades for £2.50, in 2016. It's now 2018, and I'm on my last blade. Spending another £2.50 will see me through to 2019.

What you can do now

Grow a beard! Or try an electric razor, or some other kind of depilation that doesn't use plastic.

FEMININE HYGIENE PRODUCTS

In Australia, plastic tampon applicators are known as 'beach whistles'. That tells you something, doesn't it? From our figures, 'beach whistles' make up 0.9 per cent of all litter picked up. This means that almost 1 in every 100 pieces of plastic picked up was a plastic tampon applicator.

How do they get on the beach? If you flush them, they go to the sewage-treatment works, where they usually get picked up, but when there is a lot of rain water, companies are allowed to let sewage and rainwater overflow into the sea and rivers. Anything in that – tampon applicators, sanitary towels, nappies – will get washed out too and end up on the beach, breaking down into microplastics and threatening wildlife.

Ewww!

The #2minutesolution

If you have to use plastic tampon applicators for medical reasons, then please dispose of them properly, in a waste bin. If you don't, take 2 minutes to find tampons with cardboard applicators or try applicator-free tampons. It's so simple to make this change and yet it will have a huge effect.

There are also some reusable tampon applicators on the market.

If you live in the city, it still counts so please don't think it doesn't affect you. All rivers lead to the sea.

What you can do now

Spend 2 minutes doing some reading around the subject. I am in no position to tell tampon users what to do, but suffice it to say that I know you don't have to do something simply because it's what you've always done.

YOUR DAILY WASH

Your wash could be killing the oceans.

Your washing machine is directly connected to the ocean. It might be a convoluted journey, but it is a reality. Waste water travels down the drain to a sewage plant, where it will be treated and filtered, then either put back into the water system or released via freshwater outfalls at the coast or in rivers.

The problem with this is that your clothes will release micro fibres every time you wash them. The fibres are too small to be picked up by normal filters either in your machine or in the water-treatment works. That's OK if it's made of wool, cotton or other natural fibres, but every item of clothing made from man-made fibres – nylon, acrylic, polypropylene and elastanes – sheds plastic lint each time you wash it.

A study by a group of students at the University of California Santa Barbara's Bren School of Environmental Science and Management found that, when washed in a washing machine, a fleece sheds, on average, 1.79 grams of micro fibres per wash.

The #2minutesolution

The good news is that you can do something about this. Stop wearing clothes from man-made fibres. Of course it's not that easy, or cheap, but may lead you to buy better quality clothes that will last longer.

You can also buy an in-line filter for your washing machine. Or a cheap alternative is to buy a Guppy Bag, a lint catching bag that can be used in all washing machines. It will catch a lot of the micro fibres that come off your clothes and costs around £27.

If you can't give up the gym kit, fleeces or technical clobber, there are a few simple ways to reduce their impact:

- Wash clothes less frequently.
- Wash clothes on shorter washes.
- Wash clothes at lower temperatures.

- Make sure your machine is full – clothes in full machines get agitated less and therefore shed fewer fibres.
- When you clean out your dryer put the lint in the rubbish rather than down the sink.

What you can do now

Look at the clothing labels in your wardrobe. Make a mental note of what contains man-made fibres and vow to wash them less frequently. Half as often is half the microfibres making their way to our oceans.

YOUR WEEKLY SHOP

I am incensed by food packaging. Everything is covered in plastic, from meat to vegetables, ice cream to bread. Arguably, it's for freshness and hygiene, and that does make sense, to a point, it's just that there is so much of it.

The problem is that I see the connection between what I walk past in the supermarket and what I pick up on the beach. More than 1 in every 20 items logged with the #2minutebeachclean app was food packaging of one sort or another.

Black plastic trays, used for ready meals and meat products to make them look delicious, are really hard to recycle.

Food and drinks pouches, clingfilm and film lids cannot be recycled easily either. They all have to go straight into landfill or are burnt for energy recovery. It's criminal.

The #2minutesolution

- Avoid black plastic trays. In fact, avoid as much as you can anything that can't easily be recycled.

- Buying from the butcher, fishmonger and deli counters at the supermarket will give you the chance to say you don't want plastic food trays.
- Take your own Tupperware or washed takeaway cartons to the supermarket and ask them to put your food in those. Some supermarkets are great at doing this. Others not so. Make your choices.
- If you can, avoid supermarkets. Go to the butcher, baker, greengrocer and fishmonger. You can easily reduce your plastic consumption this way. You'll have to go out of your way for sure, but it's not that hard to do.

What you can do now

Find your nearest zero-waste supermarket online. The idea is that you take your empty food packaging, fill it with whatever you want, weigh it and pay for what you buy. Then you take it home, without any waste. It might seem revolutionary now, but really it harks back to an earlier time when we used less packaging.

BUYING VEGETABLES

If only oranges, avocados, bananas, cucumbers, lemons, sweetcorn and grapefruits had their own outer casing or skin that would keep them fresh.

Oh, wait a minute ...

The more aware we become of the damage plastic is doing to our planet, the more we realise supermarkets are shoving anything and everything in plastic. You don't need an avocado in a plastic tray. You don't need bananas in a plastic bag. You don't need a cucumber wrapped in plastic. And yet it's almost impossible to avoid it. Sometimes the packaging they put it in isn't even recyclable, like most bags of salad. Again, this, in my view, is criminal.

The #2minutesolution

- Avoid all plastic if you can by tailoring your shop to what is not plastic-wrapped. At the very least, avoid plastic that can't be recycled. It's not easy but it can be done.
- Shop as usual but leave all the packaging at the till and let the supermarket know why you are doing it. The message will soon get through.
- If you can, buy local at a farmers' market or greengrocer, where they use less packaging and take fewer food miles to get to you.
- Think about joining a 'veg box' scheme.
- Grow your own. Or at least grow a few bits and pieces on your windowsills.

Herbs are easy to grow, as are chards and leaves like kale.

What you can do now

Next time you do your shop, find all the fruit and veg that ISN'T wrapped in plastic. Do you find it shocking how little there is?

WHITE GOODS AND ELECTRICAL BITS AND BOBS

The average washing machine is expected to have a lifespan of anything between 6 and 20 years. That means you could get through quite a few of them in your lifetime. It's the same for ovens, fridges, vacuum cleaners and other electrical and white goods.

Part of the problem with electrical items is that they become obsolete; either because they no longer function properly, because they are no longer efficient or compatible with other technologies or because they aren't designed to last (known as 'inbuilt obsolescence').

How many old phones or useless items of electrical equipment have we got lurking in our kitchen drawers or out in the garage? And how many items do we send to landfill during the course of our lifetime?

The #2minutesolution

- Take it to a repair café. These are places where people will repair, or help you to repair, your stuff for free, giving up their time to help you.
- It's not easy to buy well when it comes to large electricals, simply because those with the longest life span (and guarantee) are

often most expensive. But if you can, go for it. The longer the guarantee, the less likely it is to break.

- The WEEE Directive aims to reduce landfill. Taking your old electricals to your local recycling plant will give it a chance of being recycled into its raw materials, if it isn't resold to someone who needs it.
- If you feel the need to get rid of stuff that's still working, give it away to someone who needs it. Freecycle is a great way of giving your old stuff new life.

What you can do now

Find your nearest repair café at www.repaircafe.org. Find a better home for an unwanted item at www.freecycle.org. Research the recycling options in your area. Start at www.recyclenow.com.

According to Eurostat's paper on waste identified as part of the WEEE Directive (an EU directive to designate safe and responsible collection, recycling and recovery procedures for all types of electronic waste) 3,868,818 tonnes of electrical waste is produced within the EU each year. Germany leads the way with 721,870 tonnes, followed by the UK (663,100 tonnes), France (617, 401), Italy (314,210) and Spain (230,728), almost half of which are large electrical items.

HEALTH AND BEAUTY PRODUCTS

Shampoos, conditioners and all types of health and beauty products come in plastic bottles. They are often coloured and have plastic wraps and labels and lids made from different types of plastic, making them difficult for recyclers.

While I can't tell you to do without – and why should you? – I can plead with you to make smart choices by, at the very least, making sure every bottle you use is recyclable and recycled when you have finished with it.

There are lots of alternative products on the market, as well as brands like Lush that are working to reduce their plastic consumption, that will give you a more plastic-free way, so making informed choices will go some way to reducing your plastic footprint.

The #2minutesolution

- Buy products in larger bottles.
- Ditch bottled handwash in favour of soap.
- Buy shampoo in bars rather than bottles.
- Ask your favourite brands to think more ethically and sustainably.
- Research alternative natural and eco-friendly beauty brands.
- Don't buy travel-size bottles every time you go on holiday – buy refillables.
- Consider making your own beauty products.

What you can do now

Go to the bathroom cabinet. Dare to peek inside. Take 3 health and beauty products that come in plastic – shampoo, liquid soap and deodorant – and see if you can find non-plastic alternatives. Try them!

THE SCOURGE OF MICROBEADS

On 9 January 2018, the UK ban on microbeads came into force, making it illegal for manufacturers to make or sell products containing tiny pieces of plastic. Previously, these had been used in toothpaste, facial scrubs and other beauty products, often as a cheap alternative to other, less toxic exfoliating materials.

According to the parliamentary Office of Science and Technology, between 16 and 86 tonnes of plastic microbeads from facial exfoliants were washed down UK drains every year before the ban. They then make it into our oceans and waterways because they are too small to be filtered out in water-treatment plants.

The #2minutesolution

Beat The Microbead is a global app that helps you to identify products that contain microbeads. Download it and ditch any products you may still have that contain these hideous tiny plastic pieces.

What you can do now

Check your bathroom cabinet. Look for products you still have that have microbeads listed in their ingredients. These are identified as but are not limited to polyethylene (PE), polypropylene (PP), polyethylene terephthalate (PET), polymethyl methacrylate (PMMA) and nylon (PA) (source: beatthemicrobead.org). If you find it contains any of these, stop using it and bin it – don't wash it down the sink. If you can't find them in the ingredients list then you may be able to see them in the product. Failing that, scan the barcode with the app to find out.

9 MORE #2MINUTE SOLUTION WINS FOR YOUR HOME

Everywhere you look there's plastic! Here are just a few more ways to cut it out of your life.

YOUR CLOTHES

According to a survey undertaken by the UK supermarket chain Sainsbury's, in 2017,

Britons will throw out around 680,000,000 items of clothing each spring. Really? Cripes! That's a lot of out-of-fashion, unloved and excess fabric going to landfill. It's unnecessary.

Also, think about buying quality fabrics. Not only do man-made fibres shed tiny fragments of plastic when you wash them, they also never break down! So your nylon flares will live on for ever, somewhere. Aaargh!

The #2minutesolution

Forget fast fashion – where you buy something new every season and throw it out soon after – and invest your money in 'slow fashion', a concept that's being taken up by ethical brands everywhere. Buy well, buy once. Buy decent fabric.

Some technical fashion brands, like Patagonia, Aplkit and Finisterre, offer repair services to their customers, so you can keep their gear going for longer. The longer you keep it, the less you waste and the more money you save. Win-win!

And, of course, when you do have that inevitable clear-out, consider donating your old clothes to charity or a clothing bank. Don't just throw them away.

YOUR CUPPA

Bad news, I am afraid: most teabags contain plastic. It's a devastating blow to those of us who have been composting our teabags and then spreading the plastic on our gardens. Why didn't anyone own up sooner?

As for those individual coffee pods? It says a lot that a former boss of Nespresso denounced their pods as an environmental disaster in 2016.

The #2minutesolution

- Get a teapot and use loose-leaf tea. Once you are done with the leaves they can go straight on the compost heap like before – but without the plastic!
- It is now possible to buy plastic-free teabags. Check yours are plastic-free. If they aren't, change.
- Coffee pods are notoriously difficult to recycle. Choose a coffee machine that grinds fresh beans or uses loose-ground coffee instead of pods.

LIGHTING THE FIRE

Global giant Bic has sold over 30 billion disposable lighters in 160 countries since 1973. According to their website, they make 5 million lighters each day.

I find disposable lighters on almost every beach I visit. They are, sadly, ubiquitous.

The #2minutesolution

Use matches or invest in a refillable lighter. Matches contain zero plastic, and if they fall in the sea they will biodegrade. Simple.

WASHING UP

Lots of dish cloths are made of plastic. Scouring sponges are made of plastic. Microfibre cloths are made of plastic. It's all plastic! When you throw them away they will not decompose, sitting in landfill for eternity. Washing-up bottles are plastic too.

The #2minutesolution

Take a trip to a traditional ironmonger's,

where you'll find wooden scrubbing brushes and bottle washers and cotton cloths for the washing-up, as well as all kinds of traditionally made household items that don't contain plastic, like brooms, buckets and wooden clothes pegs.

Buying washing-up liquid in bulk can save money and also save plastic. Buying eco-friendly washing-up liquid online will cost you about £2.19 per litre if you buy a 15-litre refill box. If you buy a small 450ml bottle it will cost you £3.55 per litre and use 30 bottles instead of just 1.

MILK

Since *Blue Planet II*, my local milkman has reported an 80 per cent rise in business. Milk delivery is a no-brainer: you wash the reusable bottle once you've drunk your milk, leave it out for the milkman to take away and they will replace it with fresh milk in a clean bottle. In 1995, according to a Commons paper on the UK dairy industry, 45 per cent of milk was delivered to the doorstep. By 2014, it had declined to 3 per cent.

In Britain, we consume 7,028,000,000 litres of milk a year. If the average person buys milk in a 2-litre bottle, that's 3,408,580,000 plastic bottles used each year in the UK just to carry milk.

The #2minutesolution

Order milk through your milkman. Find him at www.findmeamilkman.net.

BUTTER

Butter comes in greaseproof paper, foiled paper and in plastic tubs. Plastic tubs can be recycled, of course, but greaseproof paper is easier, and if somehow it got lost in the environment it'll do little harm. Foiled paper can be difficult to recycle.

The #2minutesolution

Buy butter in paper, not plastic, and check it can be recycled. And buy it in large pats. That'll halve your waste. Easy.

BREAD

It's really difficult to buy bread in waxed paper these days. Most manufacturers have moved to plastic.

The #2minutesolution

Lots of supermarkets bake bread instore. Get it before they put it in those bags with clear plastic film on the front (it can't be recycled). Even better, if there is a bakery near you, buy loaves or rolls and take them home in your tote bag or in a paper bag. It's not rocket science, is it?

LOO ROLL

Most toilet paper comes in plastic wrapping these days ... why?

The #2minutesolution

Take a look at www.uk.whogivesacrap.org and order some recycled loo paper. It comes wrapped in paper too. Buy in bulk with a group of friends and save money.

NAPPIES

Until recently, disposable nappies were difficult to recycle because of their composite nature: they contain a range of materials. Each day in the UK, around 8 million disposable nappies are thrown away, with parents contributing around 5,000 nappies per child to that pile. Nappies make up anywhere between 2 per cent and 8 per cent of landfill volume, making them a big problem for all of us.

The #2minutesolution

Being a parent and making the right decision about nappies isn't easy. It's more than a #2minutesolution to do the research and try them out. But, finding the will to do things differently only takes a moment.

There are all kinds of answers. Some parents go for traditional washable nappies, while others go for more eco-friendly products as a better alternative to your bog-standard composite nappy. They often use sustainable materials such as bamboo- or wood-fibre based materials, making them compostable, but not home compostable.

One thing to remember is that disposable still means that it goes in the bin (and therefore to landfill), unless you choose to make special arrangements.

And don't forget those wet wipes. Never flush them, if you must use them. But it's better to use paper towels, reusable cloths or bamboo wipes.

There are schemes that will take away your disposables and recycle them – but check first that your sustainable nappy is compatible! It's a minefield.

Try www.nappicycle.co.uk for nappy recycling in the UK.

#2MINUTE-SOLUTION FOR KIDS

Now it's time for the kids to have their say. Believe it or not, you guys have buying power and influence. Imagine this (from an incredible idea by Andy Middleton, founder of groundbreaking adventure company, TYF):

Your school has 1,000 pupils.

Every pupil buys drink in a plastic bottle every day.

That's £1,000 per day, and there are 190 school days a year

That adds up to £190,000 per school year.

And it leaves 190,000 bottles to be recycled or going to landfill each year.

Or 5 tonnes of plastic.

Wow!

But how about if you all persuaded your school to get rid of plastic drinks bottles, install a water fountain and everyone used it? Then you put all that saved money together. What would you do with it?

Worth a thought.

FIZZY DRINKS

WRAP, the government's recycling project, claims the UK gets through 580,000 tonnes of drinks bottles each year, with only 281,000 of them being recycled. One in 20 items logged on beach cleans is a drinks bottle.

The trouble with drinks bottles is that not all of them are recyclable because the various bits of the bottle are made from different types of plastic. The lids, bases and plastic labels are often all made from different plastics and have to be recycled in different ways, or aren't recyclable at all.

Coloured plastics or opaque coloured plastics are also difficult to recycle. In fact, some colours can spoil batches of recycling or get rejected by recycling sorting machines.

Bottle-deposit schemes

The idea of a bottle-deposit scheme is nothing new. Back in the day, when I was a kid, we'd pay a little extra on our bottles of fizzy pop. When we were finished we'd take the empty bottles back to the shop and get our deposit back. It meant you'd never throw away a bottle and finding one on the street was like getting free money! It's one of those old ideas that doesn't sound so bonkers in the here and now.

Happily, in March 2018, after much opposition from drinks

manufacturers and campaigning from the likes of CPRE, Greenpeace, the MCS and SAS, the UK government announced the introduction of a deposit return system (DRS) for England.

This was great news.

After Germany introduced 25 cents on every single-use plastic drinks bottle in 2003, the return rates soared to 98.5 per cent (according to zero waste campaign group Zero Waste Europe).

Let's hope it gets the support it needs.

The #2minutesolution

- Does your school have a water fountain? If you don't have one, ask your teachers why not. Get a few of you together and petition your head teacher to provide one so you can fill up your water bottles.
- If you can't live without carbonated drinks, persuade your school to buy a Sodastream. They make fizzy drinks out of tap water and will soon pay for themselves, if you use non-branded syrups.

What you can do now

Think about the ways your school uses plastic. Could you cut any of it out? Talk to your teachers and tell them how they can do it and why it's important to stop using single-use plastic.

PLASTIC CUTLERY AND PLATES

It's unthinkable to me that schools should be allowed to serve your school dinners using disposable cutlery and plates. And yet some schools do. Does yours? How many pupils does your school have? Times the number of pupils by 190 and you'll see just how many pieces are getting thrown away each year.

Horrifying?

The #2minutesolution

Does your school serve your lunch with plastic cutlery? Start a petition to ban it. Get everyone to sign it and hand it in to your head teacher. Tell them why you want single-use plastic banned from school. It's your school and you have the right to say how it's run.

What you can do now

Think about ways in which you could cut down on single-use plastic at school. Cut one thing out this week. One thing next week.

BALLOONS

There is bad news about balloons. Even those that manufacturers claim are biodegradable are still deadly to wildlife. Once balloons reach a certain altitude they pop, shattering into shapes that can look very alluring to sea creatures – a lot like lunch – so they eat them. Biodegradable balloons take around 'the same time as an oak leaf' to degrade (according to manufacturers), which could be three months, by which time the poor animal that has eaten it will be dead.

Strings and ribbons entangle and kill sea birds and animals too.

The #2minutesolution

Avoid balloons, even if they tell you it's OK because they are eco-friendly. They aren't!

If your school wants to hold a balloon release, object as much as you can. Tweet, or get your parents to tweet the @2minutebeachclean and we'll get our family to send a barrage of kind but firm emails asking them to reconsider. We've done it before and it works.

What you can do now

Think of 3 ways to decorate a party without using balloons or plastic decorations. Go!

GLITTER

The news that glitter was ravaging the oceans was like cancelling Christmas for people everywhere. But glitter is readymade microplastic. It washes straight out to sea and into the path of the fish and sea life that will mistake it for food. Like microbeads in cosmetics it cannot be filtered so there is no stopping it.

The #2minutesolution

You can always try eco-friendly glitter made from cornstarch or other natural substances. But they still take time to break down – sometimes as much as 90 days – so the word is caution. That said, they are much better than plastic if you can't repress your inner glitterpuss.

What you can do now

Ask your school to cut out glitter. Tell them why. If you need help, get everyone in your class to sign a petition.

A study by
Plymouth University,
in 2016, claimed that
microplastics were found
in a third of UK-
caught fish.

#2MINUTE-SOLUTION FOR YOUR WORK PLACE

How much time do you spend at work? Too much, I expect. In that time you consume food, drinks, stationery. You may produce products and packaging or procure products to sell on. You might run a huge business, be a tiny one-man band, work in an office or spend your days on the road. Whatever it is you do you have opportunities every day to make a difference through the choices you make. If you can inspire colleagues, bosses or your staff to make small changes across the board, they could have a big effect.

YOUR DAILY COFFEE

We love our takeaway coffees. The UK gets through 2,500,000,000 coffee cups every year. That's around 79 every second.

Until recently, none of these could be recycled because of their plastic content, which meant they all went into landfill. Imagine that. Consider also the lids on the cups, which are plastic too.

Today, there are coffee shops who will take your 'disposable' cups back to recycle, and they will even take other chains' cups too.

The #2minutesolution

There is an easy #2minutesolution here. It's so simple that it's laughable, but it does mean you may have to put yourself out a bit.

Coffee chains give discounts to people who bring 'keep cups' into their stores. Some give

50p off per refill. And a lot of them will wash it out for you too.

If you have a takeaway coffee every day of your working life you'll probably have 240 per year (if you have 4 weeks holiday), which means you can save up to £120 if you take a keep cup. With the average keep cup costing around £10 that's a 'profit' of £110 for carrying one with you.

No excuses.

What you can do now

Get yourself a keep cup. Make a point of using it. Maybe even keep a spare one in your car. Buy them for all your friends. Next time you buy a coffee, ask the barista what their policy on recycling cups is. If you get a good answer, great. If not, go somewhere else.

YOUR LUNCH BOX

Takeaways are a big part of throwaway culture. In fact, you might say that they define it. We are all so busy being busy that we've no time to make sarnies any more. We rely on food made in factories, consuming tonnes and tonnes of packaging in the process.

Polystyrene containers are difficult to recycle, while things like plastic food bags may be recyclable but are often contaminated with food. Clingfilm – also typically contaminated – can clog up recycling machines, if it's recycled at all. And 'compostable' packaging usually needs to go to an industrial composter rather than be composted in your home compost bin. Check with your takeaway how you can compost their compostable cartons – and see if they will take them back if you can't compost them at home.

In Mumbai, around 200,000 home-cooked lunches are delivered to office workers every day by dabbawalars, the tiffin-carrying food delivery men. The system delivers food from home to the office – up to 50 miles away

– in time for lunch and then returns the empty tiffins back home before the end of the working day. No waste. And it's 99.9 per cent successful, with very few lunches going missing.

The #2minutesolution

Make your own sarnies and wrap them in beeswax wrapping (100 per cent eco-friendly) or tinfoil or take them in a sandwich box. You'll save money and will eat food that you know has been prepared with love. If you have no time to make your own food, take your lunchbox into work and ask your sandwich shop to put your sandwich or salad or baked potato straight in there, so creating no plastic waste.

Failing that, go to a sandwich shop where they will wrap your sarnies in greaseproof paper and a paper bag.

What you can do now

Try making your own sarnies next Monday.

PENS

The BIC Crystal Ballpoint Pen is the world's best-selling 'disposable' plastic biro. In 2006, the company celebrated selling their 100,000,000,000th pen. Bic aren't the only company making disposable pens. And it's not just biros. Felt tips, biros, highlighters, gel pens, magic markers are all single-use plastics.

You may have a few ballpoints knocking around at home, unloved, waiting to go to landfill. And they will, because, mostly, ballpoint pens are composite – they contain metal and plastic – making the materials tricky to recover. I find them on beaches, particularly the bookies' stubby pens.

But can we live without them?

The #2minutesolution

Ballpoint pens are the only thing that can be used for certain jobs like signing credit cards. They are also clean and convenient. It is still possible to buy refillable ballpoint pens. Even though they still contain plastic components, by using them you'll still stop plastic single-use pens like throwaway biros from going to landfill or the ocean.

However, buying a fountain pen will save even more plastic, especially when you refill it from an ink bottle rather than use cartridges.

What you can do now

Fountain pens can be bought for around £10 (up to hundreds). Ink costs around £5 for a bottle, making the initial outlay higher than the average ballpoint at around £2 for 10, but think of the sense of satisfaction you'll get from producing no waste whatsoever. Try it.

#2MINUTESOLUTION FOR BUSINESS OWNERS

If you own a business of any size then you are in a position of great responsibility, not only to your customers and staff, but also to the planet. By making simple changes to the way you do business you could easily do more than I ever could with this book. Talk to your designers about the end of life of your products. Get big, bold and brave recycling messages onto your labels to educate your customers. Switch to easily recyclable materials. Look at your packaging, filling and transportation.

You have the power to change the world. So let's do it.

The #2minutesolution

- In the office/work kitchen, replace plastic throwaway cups with glasses and mugs.
- Replace washing-up utensils with non-plastic alternatives.
- Consider getting a hot tap for tea-making.
- Filter tap water and remove the water cooler.
- Consider buying milk in bulk rather than in small cartons.
- In the bathroom, use dryers instead of towels.
- Start a recycling policy for as many items as you can recycle: paper, CDs, batteries, plastics, business cards, pens.
- Rethink the stationery cupboard and use cardboard, recycled paper, non-plastic card and business cards. Stop using plastic Jiffy bags or lined envelopes.
- Go bin-free.

- Could you even go printer-free?
- If you have a canteen, remove plastic cutlery, takeaway boxes, sandwich wrapping and vending machines that sell single-use plastic.

THE SURFDOME MODEL

In 2015, Surfdome, one of Europe's biggest retailers of surf apparel and hardware, decided to try to go plastic-free. In the first quarter, they removed 74 per cent of plastic from their operations by changing to cardboard boxes and gummed paper tape. This saved the equivalent of 650,000 plastic bottles.

Three years on, they have eliminated the equivalent of 2 million plastic bottles. They have been rated as one of top 14 fashion brands saving the oceans and have become a case study for the World Business Council for Sustainable

Development for reducing waste. Everyone wants to talk to them.

The #2minutesolution

Some ideas for you and your business. It's just a start but it all helps ...

- Appoint a green champion.
- Talk to suppliers.
- Think about how you can educate your customers.
- Use cardboard instead of plastic.
- Drop sticky tape in favour of gummed tape.
- Use paper bags instead of plastic bags.
- Stop using disposables in your canteen.
- Shred used cardboard boxes to reuse as packing materials.

What you can do now

Talk to your team about reducing plastic. Ask them where they use plastic and how they might get rid of it. By taking a holistic approach the change won't cost you. And the PR will be awesome!

Make a big statement to your staff: buy them all a coffee mug and ask them not to bring takeaway cups into the office. That should start the conversation ...

THE END OF THE BEGINNING

Thanks for getting to the end of the beginning. Really though, it is just the start. If you've come with me on this journey then by now you will have made a lot of very simple changes that are starting to add up to be one big change.

Take a look at yourself.

You are starting to look like the future.

You look fabulous.

Or handsome.

(Whichever you prefer.)

What you can do now

Implement a #2minutesolution from this book. Upload a picture of it to social media using the hashtag.

Maybe you have your own #2minutesolution? Share it with the world so all of us can be inspired to make more small changes that will add up to one big change.

CHECK LIST OF THINGS TO DO

How many of the following have you managed to achieve?

How hard was it?

See how they add up?

Well done.

1. Look around at the plastic in your life. How much do you see? Could you live without it? ☐
2. Think of one item you buy in single-serve or handy packs – like tissues, yoghurts or individually wrapped ☐

crackers and dips – that's about your convenience more than the planet. Cut it out. Add up the amount of packaging you'll save from going to landfill (or the ocean) after a year.

3. Go outside, into your street, a local park or open space. Set the timer on your phone to 2 minutes and pick up litter until the buzzer goes off. How much did you collect? Surprised? ☐

4. Think of the last thing you bought that came in a blister pack. Can you buy the same product without all that packaging? Next time you need it, vote for the product with the least amount of packaging. ☐

5. Go to the veg aisle at your local supermarket. Count up the number of items that could easily be sold without plastic. ☐

6. Spend a couple of minutes going through your supermarket shop. Look at the packaging. If it says 'cannot currently be recycled' don't buy it again. ☐
7. When you're out, pick up the first plastic bottle you see, take it home to recycle. ☐
8. Look out for misleading claims on packaging. Plastic is still plastic, even if it is 'degradable'. Some 'eco-friendly' packaging can still only be composted in industrial composters. ☐
9. Find out what your local authority can take for recycling and where it goes. Make a pledge to live within what they can accept. If they don't accept some materials, ask them why. ☐
10. Email your MP. Tell them you want anti-plastic legislation, polluters to pay, and your children to grow up in a plastic-free world. ☐

11. Which, 3 out of plastic bottles, straws, crisps, cotton buds, plastic bags or wet wipes could you give up tomorrow? Do it. ☐
12. Download the Refill app now. Find your nearest free refill point and use it. ☐
13. Go to your kitchen drawer. Take out a fork. Put it in your work bag. Use it. ☐
14. Go and take a look in the cupboard. Pull out the items that have 'not currently recycled' on the packaging. Seek out alternatives. ☐
15. Undertake a 2-minute review of your toiletries. If there is plastic in them, think about how you can make changes to cut it out. ☐
16. Make a Morsbag. Download the pattern, buy your labels and start making. It won't take long to make one. ☐

17. Try a safety razor. You will have to work a little harder to get a great shave, but it's not that hard. ☐
18. Spend 2 minutes doing some reading around the subject of green alternatives to tampons and pads that contain plastic or plastic applicators. ☐
19. Look at the clothing labels in your wardrobe. Make a mental note of what contains man-made fibres and vow to wash them less frequently. Or get a Guppy Bag to wash them in. ☐
20. Find your nearest zero-waste supermarket online. ☐
21. Find your nearest repair café at www.repaircafe.org. Find a better home for an unwanted item at www.freecycle.org. ☐
22. Spend a few minutes researching plastic-free beauty products. Could any of them work for you? Try it. ☐

23. Check your bathroom cabinet. Look for products you still have that have microbeads listed in their ingredients. If you find it contains them, stop using it immediately. ☐
24. Think about the ways your school uses plastic. Could you cut any of it out? Talk to your teachers and tell them you think you should find another way. Suggest alternatives. ☐
25. Ask your school to ban balloon releases and say why. Suggest other ways of celebrating. ☐
26. Ask your school to cut out glitter. Tell them why. If you need help, get everyone in your class to sign a petition. ☐
27. Try making your own sarnies next Monday. ☐
28. Try using a fountain pen, with real ink, to cut down your pen waste. ☐

29. Talk to your work colleagues and employees about reducing plastic. Ask them where they use plastic and how they might get rid of it. ☐
30. Make a big statement to your staff: buy them all a coffee mug and ask them not to bring takeaway cups into the office. ☐

WITH GRATEFUL THANKS TO:

- ☆ The #2minutebeachclean family.
- ☆ Dolly, Nicky, Tab, Alan and the #2minutebeachclean dream team for making all of this happen.
- ☆ Liz, for sharing the best beach-clean times.
- ★ Maggie and Charlie, my beach cleaning warrior mermaids.
- ☆ Adam and all at Surfdome for believing and doing.
- ☆ Neil Hembrow at Keep Britain Tidy for being the godfather of UK beach cleaning.
- ☆ Chris Hines for years and years of inspiration.
- ☆ Tim at PFD for his enthusiasm.
- ★ Laura, Sarah, Lottie and all at Ebury.

- Ado, Rachel and Jack for ALWAYS getting stuck in.
- Deb and Kim for the quite good idea about the beach-clean stations.
- Jan, Avril, Claire, Keith, Lynda, Simon, Sue, Zoe and the Bude Crew.
- The Legendary Lifeguards of the Plastic Movement.
- The Crackington Crew.
- Tracey Williams, Nat Fee, Gus Hoyt.
- The *Springwatch* team.
- Laura and Chris.
- *Blue Planet II*.
- Sinead and Becky Finn at An Taisce.
- Annabel Fitzgerald.
- Freckleville.